D1417940

DISCARD

TEAM EARTH

ECOSYSTEM ARCHITECTS
ANIMALS BUILDING INCREDIBLE STRUCTURES

BY MARTHA LONDON

CONTENT CONSULTANT
Bruce A. Eichhorst, PhD
Assistant Professor of Practice, Department of Zoology
Southern Illinois University

Core Library

An Imprint of Abdo Publishing
abdobooks.com

Cover image: Baya weavers weave nests out of
materials such as leaves.

abdobooks.com

Published by Abdo Publishing, a division of ABDO, PO Box 398166, Minneapolis, Minnesota 55439. Copyright © 2020 by Abdo Consulting Group, Inc. International copyrights reserved in all countries. No part of this book may be reproduced in any form without written permission from the publisher. Core Library™ is a trademark and logo of Abdo Publishing.

Printed in the United States of America, North Mankato, Minnesota
092019
012020

Cover Photo: Tristan Tan/Shutterstock Images
Interior Photos: Tristan Tan/Shutterstock Images, 1; Shutterstock Images, 4–5, 14, 32–33; B. Brown/Shutterstock Images, 7; Radek Borovka/Shutterstock Images, 10; Brad Sauter/Shutterstock Images, 12–13, 45; Anup Shah/Minden Pictures/Newscom, 16–17; SPK Lifestyle Stock Photo/Shutterstock Images, 21; Tony Heald/NaturePL/Science Source, 24–25, 43; Red Line Editorial, 27, 37; Johannes Dag Mayer/Shutterstock Images, 28; Dmytro Synelnychenko/iStockphoto, 35; George Dolgikh/Shutterstock Images, 39; Mirko Graul/Shutterstock Images, 40

Editor: Marie Pearson
Series Designer: Megan Ellis

Library of Congress Control Number: 2019942013

Publisher's Cataloging-in-Publication Data

Names: London, Martha, author.
Title: Ecosystem architects: animals building incredible structures / by Martha London
Other Title: animals building incredible structures
Description: Minneapolis, Minnesota : Abdo Publishing, 2020 | Series: Team earth | Includes online resources and index.
Identifiers: ISBN 9781532190995 (lib. bdg.) | ISBN 9781644943267 (pbk.) | ISBN 9781532176845 (ebook)
Subjects: LCSH: Animals--Habitations--Juvenile literature. | Lairs (Animal habitations)--Juvenile literature. | Dens (Animal habitations)--Juvenile literature. | Nest building--Juvenile literature. | Animal habitat--Juvenile literature.
Classification: DDC 591.564--dc23

CONTENTS

ARCHITECTS EVERYWHERE

The farmer was concerned. Mounds and holes were popping up in his field. His alfalfa hay was dying. His tractor wheels got stuck in the holes. Only one animal could do this damage. That animal was the pocket gopher.

Farming in Nebraska means dealing with pocket gophers. The farmer spent years trying to get rid of them. He used pesticides, which are chemicals that kill pests. He tried different poisons. But he always felt bad about killing the gophers. It wasn't their fault he had a farm field with tasty alfalfa.

Pocket gophers burrow under the ground.

PRAIRIE DOG TOWNS

Prairie dogs burrow underground like pocket gophers. They live in large groups. These groups are called towns. Most towns have several hundred prairie dogs living in the same area. However, in the early 1900s, prairie dog towns were much larger. Vernon Bailey, a scientist and explorer, stumbled across a prairie dog town in Texas in 1901. It would have been hard to miss. The town covered 25,000 square miles (64,000 sq km). Scientists estimated that 400 million prairie dogs lived in the town.

However, the alfalfa was the farmer's livelihood. Before he started trying to get rid of the gophers, he lost 40 percent of his crop. Each year, he lost fewer plants, but the gophers were still a problem. They chewed the alfalfa roots. Some of the alfalfa was pulled entirely underground.

He tried trapping the gophers. That worked better than the pesticides. Still, it didn't solve the

Humans can take steps to appreciate and coexist with animal architects.

problem entirely. After harvest, he tilled the field. That slowed the gophers down the following year. But they always came back. The gopher tunnels spread out over the whole field.

A DIFFERENT SYSTEM

This year, the farmer tilled the field as usual. Then he tried something new. He planted a border of grain around the alfalfa. The border acts as a buffer around the alfalfa. Pocket gophers have a hard time burrowing through soil with the grain's larger root system.

The farmer also planted a different kind of alfalfa. In the past, he planted alfalfa with a single taproot. When a gopher chewed on it, the plant died. This year, he used a type of alfalfa with several roots. These plants can survive if a gopher eats only part of the root.

With these changes, the farmer is hopeful. He sees fewer gopher holes. Some holes remain, but they don't spread through the whole field. His harvest won't be as good as if he trapped the gophers. But he feels better knowing he isn't killing the animals.

BUILDING ECOSYSTEMS

Although pocket gophers can damage crops, their abilities are impressive. Their tunnels stretch hundreds of feet long. Burrowing rodents such as gophers are important animals in their ecosystems. They aerate the soil. This means they loosen the soil so air can mix in. This helps plants grow. Tunnels provide homes for the gophers and many other animals.

Pocket gophers aren't the only ecosystem architects on Earth. Birds build nests. Bees build hives.

MADE IN THE SHADE

Compass termites live in northern Australia. Compass termites build wedge-shaped mounds out of soil, saliva, and dung. Each of the mounds is broad on the east-west sides. But the north-south ends are narrow. This helps control the mound's internal temperature. Early morning sun heats up the east-facing side. But when the sun is highest in the sky and hottest, the mound has very little surface for the sun to warm.

Termites build huge mounds for their size.

Beavers create huge dams across streams. From the smallest insect to larger birds, mammals, and even coral reefs, animal structures are important. Animals build homes for themselves and their families. These structures provide protection. And when animals leave their structures, other animals can take over.

Animal architects change their habitats. Sometimes they create new ecosystems. These new habitats allow other animals to thrive. Ecosystems rely on all the animals within them.

Sometimes human activity disrupts these ecosystems. When this happens, animals find new ways to make their structures. Some use human-made waste. But other animals can't survive when their habitats change.

FURTHER EVIDENCE

Chapter One discusses the importance of pocket gophers in their ecosystems. Identify the main point of this chapter and a few pieces of supporting evidence. Then go to the article at the website below. Find a quote that supports the chapter's main point. Does the quote support an existing piece of evidence in the chapter, or does it present a new piece of evidence?

POCKET GOPHERS—LITTLE-KNOWN ANIMALS PLAY IMPORTANT ROLE IN ECOSYSTEM
abdocorelibrary.com/ecosystem-architects

CHAPTER
TWO

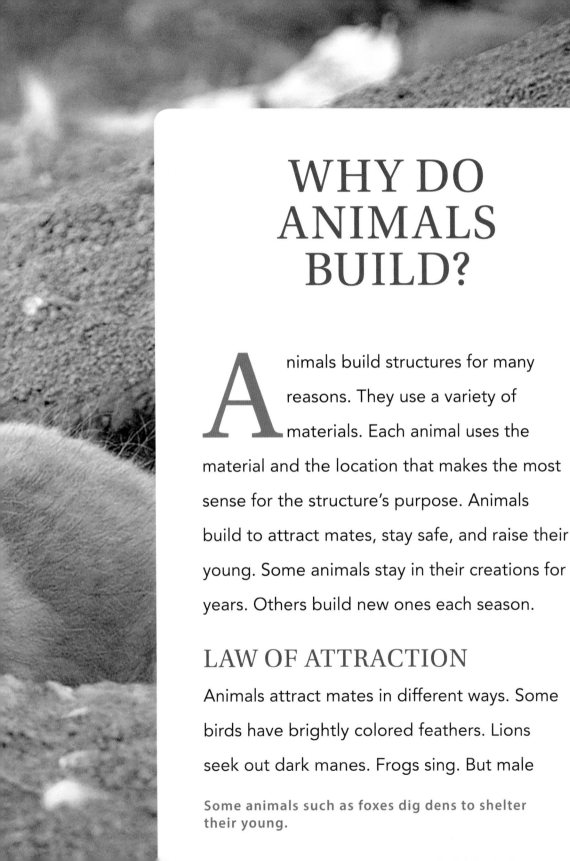

WHY DO ANIMALS BUILD?

Animals build structures for many reasons. They use a variety of materials. Each animal uses the material and the location that makes the most sense for the structure's purpose. Animals build to attract mates, stay safe, and raise their young. Some animals stay in their creations for years. Others build new ones each season.

LAW OF ATTRACTION

Animals attract mates in different ways. Some birds have brightly colored feathers. Lions seek out dark manes. Frogs sing. But male

Some animals such as foxes dig dens to shelter their young.

Bowerbirds collect found objects such as shells and small plastic items.

bowerbirds take a different approach. They design and decorate complex structures to attract females.

Bowerbirds live in Australia and New Guinea. During mating season, male bowerbirds begin building their bowers. They find a clearing where there is space to start construction. Bowerbirds find twigs to build an outer structure. These structures are called bowers. Once the bower is complete, it is time to decorate.

Bowerbirds are attracted to color. Some males paint the inside of their bowers by mixing their saliva with berry juices. Almost all bowerbirds collect objects to decorate the area around their bower. Males find flowers, berries, and shells. They bring bones and the exoskeletons, or hard, external coverings, of dead beetles. These objects are carefully laid around the clearing. Some birds will even collect bottle caps, aluminum foil, and colored glass. A male can spend weeks perfecting his bower. Once complete, he waits for a female to see his creation.

After mating is over, the female doesn't stay. The bower is never used as a nest. But the male does not always leave. He may use the same bower for his whole life. A satin bowerbird can live up to 30 years. Each mating season, a male tidies up the area, adding new materials.

SEEKING PROTECTION

One of the biggest reasons animals build is for protection. Many of the smaller animals that create structures have a lot of predators. These animals

Orangutans build nests that are out of reach
from predators.

have adapted to these threats by making their homes
underground or high in trees. These structures also keep
animals safe from extreme temperatures and weather.

Nests in trees can give animals the benefit of
height. Animals of all shapes and sizes build nests.
These animals include birds, insects, and small

mammals such as squirrels and some rats. Trees provide protection from ground-based predators and easy access to food sources.

Orangutans are some of the largest animals that build nests in the treetops. They layer branches over each other. These branches form a bed.

Orangutans build nests in trees for two reasons. The first is that the branches form a comfortable place to rest. The second reason is that trees provide protection.

In the trees, orangutans are high above the forest floor. Their natural predators include big cats such as tigers. Tigers can climb low branches. But they can rarely climb as high as the orangutans. Orangutans are also able to escape quickly from a tree nest. They leap to nearby tree branches.

Some animals do not choose the treetops. Instead, they burrow underground. This is useful for many reasons. Some holes are too small for predators to fit through. For larger

MONTEZUMA OROPENDOLA

The Montezuma oropendola is a bird that lives in Central America. It builds its nests in trees. The nests are carefully constructed. Birds choose their nest location to ensure safety from predators. The oropendola builds its nest in trees where hornets are present. The hornets keep predators and parasites away from the birds.

animals, burrows are still good protection. The holes are dark. Predators have a hard time seeing what is in the hole. This gives the animal a chance to defend itself from the predator.

One of the smallest animals to dig its home is the ant. Scientists studying ant colonies have found that ants create their tunnels in a very orderly way. One of the leading researchers in ant colony architecture is Walter Tschinkel. He makes casts of ant colonies. The casts provide a clear picture of what the inside of a colony looks like. The structures

MILLIONS OF YEARS OF DIGGING

Anthony Martin is a paleontologist. He studies dinosaurs. On a dig in Montana, Martin and his team came across a dinosaur burrow. This was one of the first times paleontologists had found evidence of burrowing dinosaurs. The find gave scientists a better idea of how some dinosaurs adapted to their environment. Until scientists found burrows, they did not know how dinosaurs survived cold temperatures. The burrows gave a possible explanation for how smaller dinosaurs lived in cold environments.

tell scientists how these tiny animals live. Tschinkel says that the structure is important. He noticed that there were different areas for specific tasks. The ants are very efficient. No two groups of ants do the same job.

Safety is important for animals. Animals build their homes to keep themselves safe from predators and weather. This becomes especially important when animals are raising young.

RAISING A FAMILY

Many animal structures have a separate place to raise young. The location and design of these structures tells scientists how these animals live. Birds known as baya weavers, like their cousins the sociable weavers, are masters at nest-building. Males build nests to attract females. Studies show that females are particular about what males they choose. Scientists found that nest location was one of the most important factors for females selecting males.

The entrance to a baya weaver's nest faces the ground.

Baya weavers live in southern and southeastern Asia. Weavers use strips of palm fronds, leaves, and vines to create their nests. Their habitat has many thorny trees. Weavers prefer to live close to water. Males build their nests with these things in mind. Food sources need to be plentiful to feed the chicks. But the nests also need to keep the chicks safe. Female baya weavers are more likely to mate with males who build safer nests. Nests made in thorny acacia trees are less likely to be raided by snakes.

Whether structures are underground or high in the trees, animals build amazing homes to keep themselves and their young safe from predators. These homes aren't just for the animals that build them. They have a big impact on the surrounding ecosystem.

STRAIGHT TO THE
SOURCE

Jane Goodall is a conservationist. After studying chimpanzee nests in 1961 and 1962, she wrote:

> *In its natural habitat, the chimpanzee . . . constructs a sleeping platform or nest on which to rest at night. . . .*
>
> *I found nests in all parts of the reserve: in the thick gallery forests of the valleys, in the trees fringing the lake, and in the more open Brachystegia woodlands of the upper slopes. The area chosen depends almost entirely on the seasonal availability of food, because the chimpanzee nests close to the trees in which he is feeding in the evening.*

Source: Jane M. Goodall. "Nest Building Behavior in the Free Ranging Chimpanzee." *Annals of the New York Academy of Sciences*, vol. 102, no. 2, 1962. pp. 455–467. Web. *Wiley*. Accessed April 10, 2019.

Consider Your Audience

Review this passage closely. Consider how you would adapt it for a different audience, such as your younger friends. Write a blog post conveying this same information for the new audience. What is the most effective way to get your point across to this audience? How does your new approach differ from the original text, and why?

ANIMAL STRUCTURES AND THE ENVIRONMENT

Animals may build their homes for themselves. But their actions affect the whole ecosystem. Other animals rely on these ecosystem architects. Rodents, amphibians, and insects use hollows that woodpeckers make in trees. Sometimes other animals even share tunnels with their creators.

Other animals build such large structures that they change the ecosystem. Beavers and termites create ecosystems with their homes.

The red ovenbird uses its nest for one season. Then other birds reuse the nest.

HOMES FOR OTHER ANIMALS

Birds build nests from many different materials. Some use leaves, grass, and sticks. Other birds create their nests using mud and their saliva. All of these nests can be used by other animals after the chicks leave.

THE MOST LOGICAL SHAPE

A beehive must be able to house the queen, workers, and larvae. It must also store all of the honey that bees make. Honeybees create their hives to maximize space. Squares are too big. Circles waste too much space. Hexagonal combs are the perfect shape. This shape uses space efficiently. The hexagon sections share sides. Bees do not have to work as hard to build combs. This means bees can conserve their energy for other activities.

One bird that creates sturdy nests is the red ovenbird. Red ovenbirds create a dome-shaped nest. Male and female pairs take turns gathering mud mixed with grass. It can take more than a month to finish the nest. Ovenbirds raise their chicks for less than three months. After the chicks leave, so do the parents.

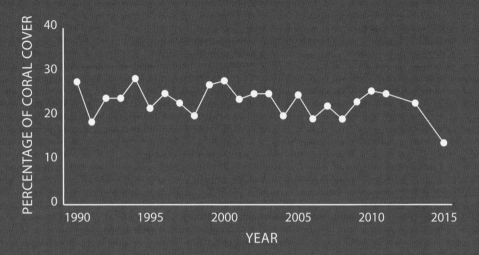

NORTHERN GREAT BARRIER REEF DECLINE

This graph shows the decline in coral reef coverage of the Great Barrier Reef. What trends do you see? After reading the chapter, can you make some conclusions about the health of the marine ecosystem? Support your claim using the graph as evidence.

This leaves a perfectly good nest. Other birds use the nest to raise their own chicks.

Birds are excellent architects on land. But under the surface of the ocean, coral is the master builder. Corals are animals. Different types of corals grow together in large systems called reefs.

Stony corals are hard. They provide the nooks and crannies that fish and crustaceans live in. Reefs are

Beavers typically build dams in streams.

also home to soft coral species and other marine animals, such as anemones. Anemones are home to clown fish. Corals grow slowly. They average less than 1 inch (2.5 cm) of growth each year. They make up only 1 percent of the ocean floor. But they support 25 percent of the ocean's marine life.

BUILDING NEW ECOSYSTEMS

Animals all play a part in their habitats. Beavers build dams. These dams block waterways. As the water backs up, it fills ponds and low-lying areas. This creates a wetland. Wetlands are important ecosystems.

Beaver dams create new wetland habitats. Many species of animals and plants live in wetlands.

Beavers are almost always building. They cut down trees. The beavers layer the branches in a stream. They pack mud around the branches to make the dam waterproof. They build a lodge in the middle of the stream. This lodge gives them a warm, dry place in the winter months. It also protects them from predators. The entrance to the lodge is underwater.

The creation of wetlands also has other benefits. Wetlands keep the world clean. As rains flood a wetland, chemicals

TAKING NOTES

Nick Weber studies the effects of beaver dams on the environment. He found that organizations could use the same technology beavers use to restore ecosystems where beavers no longer live. Weber described how scientists could take notes from the beavers to create new wetland habitats. By using flexible vegetation, such as willow and alder branches, scientists could weave the branches between posts to create a dam-like construction.

from lawns and farmlands wash into the wetlands. The grasses and mud there hold on to the chemicals. The chemicals do not enter the stream below the dam. The dam itself acts as a filter, cleaning the water. Cleaner water allows insects and fish to live in the stream.

TINY BUILDERS

Another impressive builder is very small. Cathedral termites in Africa and northern Australia build huge mounds. Some mounds stretch as high as 15 feet (5 m). Within these mounds, termites grow fungi gardens. The fungus grows in large cavities of the mound. These cavities are called galleries. The fungus breaks down undigested plant material. This is changed into material that termites can eat. The termites and fungi have a symbiotic relationship.

Termites live all over the world. Not all build impressive mounds. But they are all architects of their habitats. Scientists are just beginning to understand termites' importance in their ecosystems. Most termite mounds are short and wide. The mounds create better soil for plants to grow because termites aerate the soil. This helps the soil hold more moisture. Termite mounds are spread out. This helps plant life spread out too. As grasses grow through the African savanna, the effects of drought decrease. When rains do come, areas recover faster when termites are present. The soil has more nutrients to support plants. Scientists stress that termites do not fix climate change. Climate change affects every part of an ecosystem, even its architects.

EXPLORE ONLINE

Chapter Three discusses beaver dams. The article below goes into more detail about the dams. Does the article answer any questions you had about dams?

BEAVERS
abdocorelibrary.com/ecosystem-architects

CHAPTER
FOUR

HOW PEOPLE AFFECT ANIMAL STRUCTURES

Humans must share Earth with animals and plants. Humans are spreading further into habitats that animals need. Construction, factories, and other aspects contribute to climate change. These changes affect how and where animals build their structures.

CLIMATE CHANGE

Climate change is the process of human action contributing to warming the planet.

Expanding human development means that people are affecting animal habitats more and more.

When people build large factories, and burn fossil fuels, gasses are released into the atmosphere. These gasses trap the sun's heat around Earth. In the past century, Earth warmed 1 degree Fahrenheit (0.6°C). The ocean warmed 0.18 degrees Fahrenheit (0.1°C). This might not seem dangerous. It would be difficult for a human to notice such a small change in water temperature. However, coral is very sensitive.

When one part of an ecosystem dies or becomes less plentiful, the whole ecosystem changes. Coral supports a wide diversity of fish and other marine life. This diversity supports larger predators. Coral reefs also act as natural ocean walls. A reef slows large waves during storms. This prevents shore erosion, protecting houses during storms. When the temperature of the ocean changes, corals begin to bleach, or turn white. Bleaching doesn't mean the coral dies. But the coral is more likely to get infections because it is stressed. These infections cause coral to die.

If too much coral dies from bleaching, important underwater and shoreline ecosystems will be damaged.

The rise in global temperature affects land animals too. One group of architects that climate change seems to be affecting is bees. As the temperature rises, some bees have moved out of their typical habitats toward cooler areas. Other bees seem less able to adjust. Their populations are decreasing. The United States relies on bees to pollinate crops. Researchers estimate that bees provide more than $15 billion in service by pollinating. A loss of bees would make it more expensive to grow crops.

HABITAT LOSS

Human activity takes up a lot of space. People build roads, houses, and huge warehouses. All of this takes away animal habitat. Farms need wide-open spaces.

UNDER THE SURFACE

Mark Eakin works with the National Oceanic and Atmospheric Association (NOAA). Many scientists are studying the death of coral reefs. Eakin explains that the damage is vast. And it is lasting much longer than a single event. While bleaching can happen during a single season, scientists are finding that bleaching is happening across multiple seasons. These corals are old. Some corals have been growing for 500 years. Eakin stresses that it is impossible to grow back something that large in a short time.

DEFORESTATION IN CAMBODIA

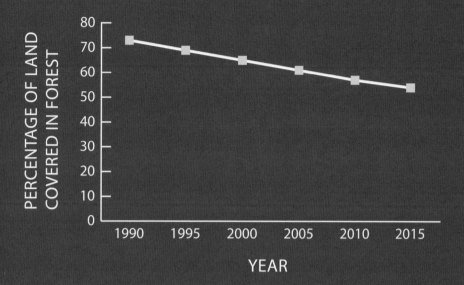

This graph shows the percentage of land in Cambodia covered in forest from 1990 to 2015. Based on the information about deforestation in Chapter Four, what do you think some of the consequences of deforestation in Cambodia are? Use this graph as supporting evidence.

People cut down large areas of forest to make room for grazing or planting.

Sometimes animals can work around human activity. But other times, that isn't possible. Deforestation and pesticide use have consequences for a whole ecosystem. Bees, birds, termites, and rodents all find ways to use human structures to build their homes.

This often causes problems for people. Termites damage homes. Bees need to be relocated so people don't get stung. Rodents tear holes in people's food. Rat and mouse droppings litter drawers and cabinets.

For larger animals, taking shelter in a person's home isn't an option. Deforestation takes shelter and food sources away from animals. Industry pollutes waterways. Beavers may leave streams they once dammed. If they cannot leave, they often die. Ecosystems are able to withstand many dramatic changes. However, changes

SOCIABLE WEAVERS

Sociable weavers are birds found in southern Africa. Weavers build the largest nests of any bird. Colonies of sociable weavers can be as large as several hundred members. Their nest acts like an apartment building. They build a massive nest of grass and twigs. The birds live in a series of connected chambers inside the nest. Human development has reached sociable weaver habitats. Some colonies now build their nests on telephone and electric poles.

Some animals, such as rats, have made use of human structures.

in forest cover and climate change happen faster than most ecosystems can handle.

Even in areas where animal populations seem to be stable, there are issues. Agriculture is important. Farmers need to harvest as many crops as possible to make a living. They use pesticides to keep rodents and insects away from their crops. Pesticides affect all insects, not just the bad ones. Bees, for instance, are needed to pollinate the crops. But pesticide use harms and kills them.

Pesticides are not the only threat to bees. Varroa mites are from Asia. They are tiny and can travel easily.

Sometimes several varroa mites feed on a single bee pupa.

Humans have accidentally brought them all over the world. The mite is very harmful to bees. Varroa mites eat young bees. An infestation of mites can kill a hive.

It is possible for people and animals to live together. Engineers in Africa are changing how they design telephone and electrical poles to better withstand the weight of sociable weaver nests. Beekeepers rescue and relocate hives from houses. But it will always be up to people to make choices that benefit animal ecosystems. People are the largest ecosystem architects on the planet. It is important to design with the rest of our habitat in mind.

STRAIGHT TO THE
SOURCE

In 2017, the Center for Biological Diversity released a statement on North American bee populations:

> *The widespread decline of European honeybees has been well documented in recent years. But until now much less has been revealed about the 4,337 native bee species in North America and Hawaii. These mostly solitary, ground-nesting bees play a crucial ecological role by pollinating wild plants and provide more than $3 billion in fruit-pollination services each year in the United States. . . .*
>
> *[A] growing body of research has revealed that more than 40 percent of insect pollinators are highly threatened globally, including many of the native bees critical to unprompted crop and wildflower pollination across the United States.*

> Source: "Hundreds of Native Bee Species Sliding Toward Extinction." *Center for Biological Diversity*. Center for Biological Diversity, March 1, 2017. Web. Accessed April 4, 2019.

What's the Big Idea?

Read the primary source text carefully. Determine the main idea and explain how it is supported by details. Name two or three of those supporting details in your answer.

FAST FACTS

- Ecosystem architects come in all shapes and sizes.

- Animals build structures for shelter, for safety, and to raise young.

- Animal structures make a big impact on the environment.

- Some animals use abandoned nests and tunnels rather than building their own.

- Male bowerbirds create intricate bowers to attract a mate.

- Beaver dams create wetland habitats used by many different animals.

- Termites aerate soil, which helps different species of plants grow.

- Human activity is hurting ecosystem architects.

- Coral bleaching is killing reefs all over the world, which is destroying habitats for many species of marine life.

- When architects disappear from an ecosystem, the whole area is affected.

STOP AND
THINK

Tell the Tale

Chapter One of this book discusses a farmer's experience trying to clear his field of gophers. Imagine you are in the farmer's shoes. Write 200 words about the methods you've tried and what effect the gophers are having on your field. How could you avoid harming the ecosystem while still growing crops?

Surprise Me

Chapter Three discusses the effects of animal architecture on the environment. After reading this book, what two or three facts about these effects did you find most surprising? Write a few sentences about each fact. Why did you find each fact surprising?

Dig Deeper

After reading this book, what questions do you still have about ecosystem architects? With an adult's help, find a few reliable sources that can help you answer your questions. Write a paragraph about what you learned.

Say What?

Studying ecosystems and animal behavior can mean
learning a lot of new vocabulary. Find five words in this
book you've never heard before. Use a dictionary to
find out what they mean. Then write the meanings
in your own words, and use each word in a
new sentence.

GLOSSARY

adapt
to change so one is able to live in an environment

cast
a replica of a formation made from a material such as metal or plaster

conservationist
a person who works to protect plants and animals

ecosystem
the living things in an area and the ways in which they interact with each other and their environment

erosion
the process of wearing away landforms through wind and water movement

fossil fuel
a type of fuel made from the decomposed remains of dinosaurs from millions of years ago

parasite
an organism that feeds off the life of another organism

symbiotic
having to do with living together in a way where each party helps the other

taproot
a main, large root that grows straight down into the ground

till
to loosen the soil and prepare it for planting

ONLINE
RESOURCES

To learn more about ecosystem architects, visit our free resource websites below.

Visit **abdocorelibrary.com** or scan this QR code for free Common Core resources for teachers and students, including vetted activities, multimedia, and booklinks, for deeper subject comprehension.

Visit **abdobooklinks.com** or scan this QR code for free additional online weblinks for further learning. These links are routinely monitored and updated to provide the most current information available.

LEARN
MORE

Hyde, Natalie. *Great Barrier Reef Research Journal*. New York: Crabtree, 2018.

Wassall, Erika. *Bees Matter*. Minneapolis, MN: Abdo Publishing, 2016.

INDEX

About the Author

Martha London is a writer and educator. She lives in the Twin
Cities in Minnesota. When she isn't writing, you can find her
hiking in the woods.